CONTENTS

T3-AIB-950

INTRODUCTION

On March 11, 2011, a 9.0 magnitude earthquake struck Japan near the Tokyo Electric Power Company's (TEPCO) Fukushima Daiichi Nuclear Power Plant. Its nuclear reactors were designed to shut down during an earthquake. But the earthquake triggered a tsunami. A wall of water up to 30 feet (9 meters) high inundated the coast of Japan, traveling as far as 6.2 miles (10 kilometers) inland. (Tsunami waves reached heights of up to 133 feet [40.5 m] in Miyako.) The earthquake and tsunami killed nearly sixteen thousand people, with many more missing and injured. It also flooded the power plant. The water rendered the backup generators ineffective. The reactors heated up, resulting in a meltdown; at least three nuclear reactors suffered explosions because of hydrogen gas buildup. Radioactive material leaked into the air and water. Hundreds of thousands of people were forced to evacuate a more than 12-mile (20 km) area around the plant.

Dangerous radiation levels went beyond the 12-mile zone. Radioactive material was found in food and tap water 140 miles (225 km) away. The cleanup area was as large as New Jersey and involved scrubbing buildings and removing tons of soil. Many people said they would never go home. Others have remained in contaminated zones. Some only allow their children outside to walk to school. No immediate deaths occurred as a result of the power plant meltdown. Some have said this shows how safe nuclear energy really is. Others say the long-term health effects of radiation exposure from the meltdown remain to be seen.

A little girl is screened for radiation exposure on March 16, 2011, in Fukushima Prefecture, Japan. Days earlier, an earthquake and tsunami damaged the Fukushima Daiichi Nuclear Power Plant.

Radioactivity happens when the nucleus of an unstable atom disintegrates—either naturally or when manipulated by scientists (as in nuclear fission). An atom is the smallest unit of an element. At the center, it has a nucleus containing one or more protons and usually one or more neutrons. One or more electrons revolve around the nucleus; the number of electrons equals the number of protons in an electrically neutral atom. When the nucleus disintegrates, it gives off particles, such as alpha or beta particles. An alpha particle consists of two pro-tons and two neutrons. A beta particle consists of either an electron or a positron. The nucleus can also give off energy in the form of a ray, such as a gamma ray or X-ray. These particles and rays are known as radiant energy or radiation.

Elements have various isotopes. That means they have the same properties but different numbers of neutrons. Some isotopes may be radioactive and others not. For instance, hydrogen is the only element with no neutrons. It is not radio-active. Its isotope tritium has two neutrons. It is radioactive.

Everybody is exposed to some radiation through nature and common products such as cell phones. It is absorbed through the skin, inhaled, eaten, or drank. Humans can tolerate small amounts of radiation, but too much can cause cancer and other health problems. That's why U.S. government agencies such as the Environmental Protection Agency (EPA), the Department of Energy (DOE), the Nuclear Regulatory Commission (NRC), and the Department of Homeland Security monitor and regulate radioactive material. Still, contamination sometimes occurs or lingers as a result of past carelessness.

Radioactive contamination can be caused by nuclear weapons, dirty bombs, power plant accidents, lost radioactive materials, and the improper disposal of nuclear waste. After contamination, radioactive material can settle into dust particles. It can seep into soil or water, where it's absorbed by plants and ingested by animals. People may eat those plants or animals or drink milk from them. They may also inhale the radioactive material. If the radioactive atoms have a short half-life, they will be released quickly from your body. (Half-life is the time it takes an amount of substance undergoing decay to decrease by half). But some radioactive atoms survive for many years, emitting radiation that can damage your cells. (Cells are building blocks of your body. Each cell contains all of your DNA, the instructions for building and operating your body.)

So why are these potentially dangerous elements in use at all? It's because radioactive elements are helpful in some ways. They create power, detect and treat illnesses, and are useful in industry. In this book, you'll learn about how radioactive elements are used, as well as your risk of contamination.

1 RADIOACTIVITY: A CURE AND A CURSE

Marie and Pierre Curie discovered radium, a radioactive element. They saw that it behaved differently from stable elements. As it released particles or rays, the atoms decayed. By the 1920s, radium, in the form of radon gas, was being used to treat cancer—an exciting medical breakthrough. However, scientists and others who worked with radioactive elements experienced health problems. Eventually they realized that while radioactivity kills cancer cells, it also damages healthy cells, causing cancer. Both Curies experienced health problems. While Pierre died in a road accident, Marie died of aplastic anemia at age sixty-seven as a result of working with radium. Radioactive elements are still used in medicine, but workers now know to handle them with care.

If you've ever had an X-ray, you've experienced radio-activity. An X-ray can determine whether a bone is broken. X-rays pass through soft tissue easily, but they don't pass

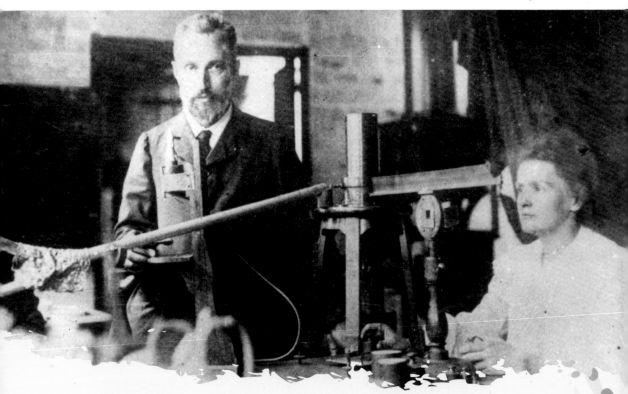

easily through bone. This causes bones to appear white on X-rays, allowing doctors to see fractures. Through this procedure, you are exposed to a small amount of radiation, which varies depending on the body part being observed. A plain chest X-ray exposes a person to the amount of radiation usually absorbed by a living thing over ten days. This is because some radiation, known as background radiation, occurs naturally.

We absorb cosmic rays from outer space. These are nuclear particles that fall to Earth when particles from outer space bombard

Marie and Pierre Curie discovered the radioactive element radium in the early 1900s. Scientists learned that while radiation could be used to treat cancer, it could also cause cancer and other health problems.

atoms in Earth's atmosphere. Common appliances and gadgets, including microwaves, televisions, and cell phones, emit small amounts of radiation. We're also exposed to radon in our homes. Radon is a radioactive element that occurs naturally in the soil and seeps through a home's foundation. The amount varies by region of the country and from plot to plot. For this reason, everyone should test their homes for radon levels.

Your risk of developing fatal cancer from an X-ray of your ankle is negligible—less than one in a million. Your risk of complications from an untreated broken bone is much greater.

A patient prepares for a CAT scan. A chest CAT scan exposes a patient to the radiation usually encountered over several years. However, it can help diagnose serious illnesses, so the benefits outweigh the risks.

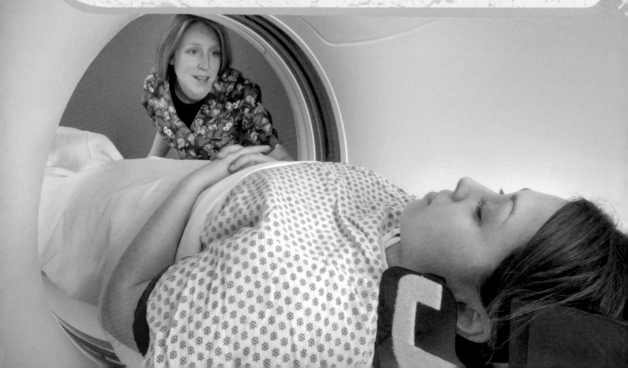

Other X-ray procedures result in greater exposure. A CAT (computer axial tomography) scan consists of multiple X-rays. A CAT scan of the abdomen and pelvis would expose you to the amount of radiation normally absorbed in five to ten years. This would increase your risk of fatal cancer from one in ten thousand to one in five hundred. Even that figure is small, however, compared to the one in five chance that everyone has of eventually dying of cancer. Also, a CAT scan could help doctors diagnose a serious illness, so the benefit may outweigh the risk. Doctors consider how many X-rays and CAT scans you've had in your lifetime. They use this information to weigh the risks and benefits.

Doctors use radiation not only to diagnose but to treat illness. Cancer is any number of diseases in which abnormal cell growth may cause tumors that disrupt body functions. (Leukemia is a unique form of cancer in which abnormal white blood cell growth disrupts the production of red blood cells). In radiation therapy, doctors use X-rays, gamma rays, and charged particles to damage the DNA of cancer cells. This causes the cells to die. Unfortunately, the radiation also damages healthy cells. Doctors limit radiation to what that part of the body can handle. However, patients typically feel side effects, such as extreme fatigue and nausea.

Radiation can be transmitted to your body through a machine or taken internally, through injections or ingestion. Patients who receive external radiation therapy do not become

THE RADIOACTIVE KID

When Taylor Wilson was fourteen, he built a nuclear reactor. This allowed him to turn atoms into radioactive isotopes. Taylor, who attends a school for gifted students in New Mexico, received a special university lab in which to safely conduct his work. He must wear a badge to measure his radiation exposure. A safety inspector visits the lab.

Far from seeing Taylor as a safety threat, government officials believe Taylor's work could make America safer. He's received offers of funding from the U.S Department of Homeland Security and job offers from defense contractors. The young scientist received second prize at the Intel Science and Engineering Fair for inventing a water-based nuclear weapons detector to test cargo coming into America.

radioactive. Patients who receive radiation internally will give off radiation. For the short time they are radioactive, they are discouraged from having contact with people, particularly children.

If you go into the medical or research field, you may handle radioactive materials or conduct radiation therapy. Be sure to follow all safety procedures. However, the medical examples above are less to show your risk of exposure and more to show that radioactive elements can be helpful when used carefully.

Medicine isn't the only industry that uses radioactive materials. Radiation is used to kill bacteria, germs, and insects in food, medical supplies, and wood flooring. It does not make the product radioactive (just as, after an X-ray, you are not radioactive). Well logging uses radioactive material to find oil and gas underground. Gauging devices containing radioactive materials measure the thickness of products—including sheet metal, newspapers, and napkins—as they are manufactured.

For these uses, the radioactive material is safely contained in metal devices called sealed sources. Sealed sources are heavily regulated. However, some fall out of regulation, becoming "orphan sources." If the orphan sources are disposed of improperly, they can be recycled into scrap metal. Then the recycled metal may become radioactive. This happened in 1997 at Royal Green Scrap Metal Recycling. A sealed source containing radioactive americium was recycled. It was detected during shipping. The EPA, NRC, and DOE helped the company find the source. The radioactive material had already contaminated the aluminum scrap metal, which had to be safely contained.

Occasionally, radioactive metal gets farther. In 2012, a shipment of brushed metal tissue boxes heading for Bed, Bath & Beyond set off alarms at a truck stop. They were found to contain cobalt 60, which can cause cancer. An NRC spokesperson said that exposure for thirty minutes a day for one year would have been the equivalent of two chest CAT scans.

The shipping of radioactive materials is regulated. Trucks must be able to contain radioactive substances even in the case of an accident. Cargo that is not supposed to be radioactive is tested to ensure that nuclear materials aren't being shipped illegally. The 9/11 Commission made it a goal to scan all cargo for nuclear materials. In addition to detecting weapons, these scans can also catch apparent accidents such as the tissue boxes shipment.

Our Radioactive Past

Today, while mistakes occur, monitoring systems and regulations tend to keep radioactive materials contained. In the past, radioactive materials were often handled and disposed of haphazardly. Improper handling that occurred decades ago can still haunt neighborhoods and towns today.

In Concord, Massachusetts, the company Nuclear Metals manufactured depleted-uranium-tipped munitions for the military. They disposed of depleted uranium in an unlined basin. This allowed the radioactive waste to seep into the soil and possibly the groundwater. The old buildings are also contaminated. Because they are in disrepair, radioactive material is leaking out. The property is now a high priority Superfund site. Superfund sites are highly toxic cleanup zones, coordinated by the EPA.

In many cases, the U.S. government itself was to blame for careless disposal of nuclear waste. In California, one high-priority Superfund site is the Laboratory for Energy-Related Health Research, which was run by the DOE. Here, from the 1950s

to the 1980s dogs were exposed to radiation to determine the health effects of long-term exposure. The soil beneath the dog pens is contaminated from dog feces, which was radioactive. Waste stored on-site also seeped into the soil. The DOE is now cleaning up the property, which involves replacing radioactive soil with clean soil.

Sometimes, radioactive contamination goes unnoticed for many years. From the 1890s to the 1940s, Welsbach & General

A strontium 90 tank is removed from the Laboratory for Energy Related Health in California as part of a Superfund cleanup effort. The Department of Energy once operated the laboratory and is now decontaminating it.

manufactured gas lamp mantels. These were cloths that formed a dome outside the burning fuel. To make the lamps glow brighter, the company coated the mantels with thorium extract. Thorium is radioactive. Its half-life is 14.05 billion years. Those exposed to it have an increased risk of cancer. The company allowed thorium to seep into the soil around the plant. When electricity led to the downfall of the gas lamp industry, the company went under, leaving a legacy of radioactive soil.

Not knowing about the contamination, residents and other businesses moved in. In the 1990s, the New Jersey Department of Environmental Protection investigated sites in Camden and Gloucester City. They discovered the contamination—which included levels of radon that caused concern—and began cleaning it up. Some of the cleanup literally occurred in people's backyards.

2 GREAT POWER AND GREAT DESTRUCTION

The work of many physicists culminated in the discovery of nuclear fission by Lise Meitner, Otto Hahn, and Fritz Strassmann. They found that when a uranium atom absorbed a neutron, it split in two. This released great energy—ten million times more than an atom of carbon when coal is burned. The implications were clear early on: this could create great power as an energy source replacing coal and oil, or great destruction in the form of a bomb.

The research for both went hand in hand; the DOE developed and tested nuclear bombs and researched nuclear energy. This is one reason why some people have been against nuclear power. For instance, former president Jimmy Carter said on May 13, 1976, "There is the fearsome prospect that the spread of nuclear reactors will mean the spread of nuclear weapons to many nations." Carter said that nuclear power would lead to more uranium enrichment facilities, which could lead to nuclear weapons.

While it's true that nuclear power and nuclear weapons programs are closely tied together through research and materials, it's important to distinguish between a nuclear power plant and a nuclear bomb in terms of the potential for harm. Nuclear bombs are designed to release all their energy at once, whereas nuclear power plants produce a steady flow of energy. Even in the case of a meltdown, nuclear power plants do not cause anywhere near the destruction that bombs do.

Today's nuclear bombs can cause six hundred times more destruction than early models. (The United States has now dismantled its most powerful bombs as part of its commitment to nuclear disarmament). And early nuclear bombs—known as atomic and hydrogen bombs—were extremely destructive. The first nuclear bomb used in warfare was dropped on Hiroshima, Japan, on August 6, 1945, by the United States in an effort to bring about the end of World War II. According to many sources, 140,000 people died from the immediate impact and effects of radiation. An estimated 100,000 more later died from radioactive contamination in the air, water, and food, according to a March 16, 2011, *Los Angeles Times* article. Three days later, a second atomic bomb was dropped on Nagasaki, Japan, killing 74,000 and injuring many more. (Japan surrendered on August 15, 1945). Even after the initial death toll, many people continued to suffer from illnesses caused by radiation. These included cancers, hepatitis, strokes, and heart attacks. In fact, much of what we know about the harmful effects of radiation is based on studies of Hiroshima and Nagasaki survivors.

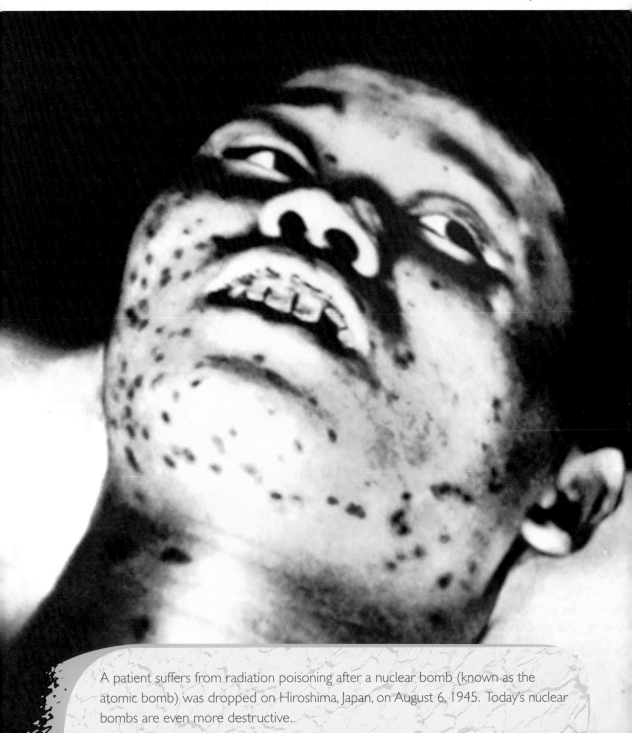

A patient suffers from radiation poisoning after a nuclear bomb (known as the atomic bomb) was dropped on Hiroshima, Japan, on August 6, 1945. Today's nuclear bombs are even more destructive.

Soon after World War II ended, the Cold War began. Though allies in World War II, the United States and the Soviet Union (now Russia) became enemies afterward. A nuclear arms race began. Other nations also built nuclear bombs.

Thankfully, these increasingly powerful bombs were not unleashed as weapons after World War II. However, they were tested, and this released radioactivity into the environment. For instance, from 1951 to 1963 the United States tested about one hundred nuclear bombs above ground. Winds blew radioactive fallout thousands of miles, contaminating the air, water, and crops. People who were children growing up during this time are at risk for developing thyroid cancer. A ban on nuclear weapons testing was signed by most nations in 1996. By then the Cold War had ended. Today, much of the work done by the DOE involves cleanup from nuclear weapons testing during the Cold War.

The Treaty of Non-Proliferation of Nuclear Weapons went into effect in 1970. One hundred and ninety countries have signed the treaty. Signees agree not to acquire or build nuclear weapons. Nations that already had nuclear weapons agreed to share technology for nuclear energy and work toward disarmament. Still, some countries, such as North Korea, have continued to build and test nuclear weapons.

After the terrorist attacks of September 11, 2001, dirty bombs became another concern. Dirty bombs are traditional explosives (not nuclear bombs) that contain radioactive materials. When these bombs explode, radioactive particles are dispersed. Dirty bombs are easier to make than nuclear bombs, so terrorist groups could devise such attacks without having advanced nuclear weapons programs. This is one reason why sealed sources and other radioactive materials are monitored so closely.

Power Plants

Like the energy in the atomic bomb, nuclear power is generated through nuclear fission. Opposite to a nuclear bomb, fission at a power plant occurs in a slow and steady manner. This is considered green energy. Nuclear power plants do not release pollutants into the air. They don't release carbon dioxide into the atmosphere, as coal, oil, and gasoline do. This greenhouse gas is believed to be a major contributor to global warming. Americans consider global warming to be the most pressing environmental issue today, according to an October 31, 2006, survey by the Massachusetts Institute of Technology (MIT).

Living near a nuclear power plant only minutely increases your radiation exposure. (Less than watching television and less than living near a coal-fired power plant, according to the EPA). However, a meltdown or leak can allow radioactive contamination into the environment. Chernobyl is the worst nuclear power plant meltdown in history. In 1986, the Chernobyl Power Plant in what was then the Soviet Union was home to four nuclear reactors. (A nuclear reactor is a device that initiates and maintains steady nuclear fission for the production of power). Workers and their families lived in nearby Pripyat. On April 26, a surge of power led to a nuclear meltdown. The disaster, a combination of poor design and human error, caused radioactive material to travel across Europe and Asia.

Workers at the power plant, firefighters who responded to the emergency, and the cleanup crew were hardest hit. Two

A mother cradles her daughter, who was terminally ill with leukemia after the Chernobyl disaster. It is believed that the radiation emitted by the 1986 crisis caused much of the illness that occurred afterward.

hundred thirty-seven people were diagnosed with acute radiation syndrome (ARS), and twenty-eight died that year, according to the World Health Organization (WHO). From Pripyat and surrounding areas, 135,000 people were evacuated, and 230,000 people were later relocated. A major health effect for these people was mental anguish. Residents lost their social networks, financial stability, and homes. They worried about their health and were treated as contaminated people in their new homes.

Over six thousand children in Russia, Belarus, and the Ukraine have been diagnosed with thyroid cancer. Many of these cancers were most likely caused by radiation exposure shortly after the incident. This form of cancer is treatable and even curable where surgery is possible. In incurable cases, a lifetime of medication is required. Leukemia and other forms of cancer are also believed to have resulted from the disaster, although the United Nations Scientific Committee on the Effects of Atomic Radiation states that the incidence of leukemia has not been elevated. WHO estimates the disaster will be responsible for 4,000 cancer deaths among workers, evacuees, and the 270,000 people who live in the most contaminated zones.

Five million people continue to live in contaminated zones in Russia, Belarus, and the Ukraine. Residents have inhaled radioactive dust, eaten food grown in contaminated soil, and drank milk produced by cows feeding on contaminated grass. WHO predicts an additional five thousand cancer deaths due to radiation exposure in these areas.

RADIOACTIVE MYSTERY

Karen Silkwood was an employee at the Kerr-McGee Nuclear Power Plant in Oklahoma. She was gathering information for the union to show that the plant was lax on safety. One day in 1974, she tested herself and found that she had been contaminated with radioactive material. Other tests, including at the National Laboratory at Los Alamos, confirmed this. Her apartment was also shown to be radioactive. The cause of the contamination was never fully explained. One night, she left a union meeting. Alone in her car, she was in a single car accident and died. There was suspicion that she was intentionally run off the road. Her family was rewarded $1.3 million from Kerr-McGee in a legal settlement related to the radioactive contamination. The plant closed soon after.

The disaster in Japan is the second worst nuclear meltdown. The world had learned from mistakes at Chernobyl. The Fukushima Power Plant was better built, and much less radiation was released than at Chernobyl. Even workers at the Fukushima plant didn't suffer from ARS. The Japanese government also responded well to the disaster, and its sound advice protected residents from radiation exposure. According to a March 2, 2012, *Scientific American* article, it's possible that no deaths will occur from the radiation. Toxic chemicals spilled during the earthquake and tsunami may pose a greater risk. However, as after Chernobyl, the heartbreak that comes from leaving home, and the stress of possible exposure (especially in a country that endured two nuclear bombs), are real health problems.

In the United States, the worst nuclear accident was at Three-Mile Island in Pennsylvania in 1979. A partial meltdown occurred because of a failure in the cooling system. No deaths occurred, and radioactive contamination was minimal. Still, the incident led to wariness of nuclear power. Within four years, fifty nuclear power plant construction plans were cancelled.

Aside from meltdowns, leaks sometimes occur. According to a June 21, 2011, Associated Press investigation, tritium leaks have occurred at forty-eight of sixty-five U.S power plants. The report found that most leaks remained on the plant's property; none had been found to have contaminated public water.

Finally, there is the challenge of disposing of radioactive waste from power plants and nuclear weapons research facilities. Nuclear waste—sixty-five thousand tons in all, according to CNN Money—is currently stored on-site at power plants throughout the United States. The July 11, 2011, article cautioned that if a terrorist attack or environmental disaster occurred, the waste could be released into the environment. The U.S. government was required by law to store this waste underground beginning in 1998. It was going to do this inside Yucca Mountain in Nevada. The site is dry and is set on uninhabited government land. But the people of Nevada have been opposed to this. Other sites are under consideration, but billions of dollars have already been invested in Yucca Mountain.

Three-Mile Island is the site of the worst U.S. nuclear disaster in history. In 1979, a partial meltdown occurred. However, the resulting radiation contamination was minimal.

Accidents at nuclear power plants have occurred. When they escalate into full disasters, sickness and death can occur, too. Relocation caused by a disaster can be heartbreaking, and cleanup difficult and costly. Chernobyl is an example of a meltdown at its worst. Now, power plants are better built, governments better prepared, and employees better trained. Experts say that today fear is disproportional to danger when it comes to power plants. Interviewed by NPR on March 22, 2011, Dr. Robert Dupont, an expert on fear and nuclear crises, said that while thousands died in the Fukushima earthquake and tsunami, fear was directed toward the nuclear meltdown. He said that nuclear power is still an unfamiliar idea, and people associate it with the devastation of nuclear bombs. Melanie Windridge, a nuclear fusion researcher wrote in an April 4, 2011, op-ed in the *Guardian* that other forms of energy are much more dangerous than nuclear power. For instance, she said that air pollution from coal-fired power plants causes one hundred thousand deaths per year worldwide, whereas the Fukushima meltdown will likely result in no deaths. She said that renewable energy may not meet the needs of Earth's growing population, and in terms of global warming, nuclear energy is much greener than coal or oil. She said that while radioactive contamination can't be discounted, fear shouldn't dominate the discussion of nuclear power.

3 RADIOACTIVE ELEMENTS AND YOUR HEALTH

Radiation affects your health because the energy strips electrons from atoms and breaks chemical bonds. This disrupts the natural processes controlling cell growth. Abnormal cell growth—cancer—may occur. Even as the body tries to repair itself, further abnormal cell growth can occur. DNA, the instructions for building proteins, which build and regulate your body and its functions, can also mutate because of radiation exposure, causing further health problems. Radiation is especially damaging to growing children and teens because their bodies absorb and metabolize substances differently and because they are more likely to develop certain cancers from such an exposure. More cell division provides more opportunity for abnormal cell growth.

Radioactive exposure is considered either acute or chronic. Acute refers to high-level exposure over a short period of time. Exposure can cause acute radiation syndrome. In the event of a nuclear bomb, many people suffer from ARS. In the case of a power plant disaster, workers at the

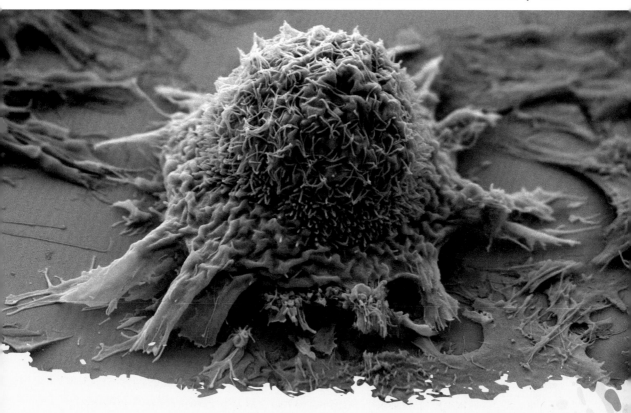

plant and first responders may develop ARS. Patients undergoing radiation treatment for cancer can also show symptoms of
ARS, including nausea, vomiting, diarrhea, fatigue, hair loss, bleeding, skin damage similar to a sunburn, and swelling of the mouth
and throat. However, doctors control the amount of radiation,
unlike cases of accidental exposure. In those cases, the patient
may appear to get better but then suffer from seizures or go into
a coma. Some people die within days or weeks due to organ
failure and damage to bone marrow, leading to infections and

Health problems, such as cancer, are clustered around areas with radioactive
contamination. Pictured here is a single lung cancer cell magnified 3,500 times.

internal bleeding. Others recover, though they may suffer from long-term health effects. Exposure is measured in rem (roentgen equivalent man). Five to fifty rem will cause only changes in blood chemistry. Fifty to two hundred will cause illness but is rarely fatal. Two hundred to one thousand causes serious illness, while doses of more than one thousand rem are almost always fatal.

The health effects of chronic exposure to radiation are less known. Scientists working with radioactive material in the early 1900s began keeping track of unusual health problems they had. But most studies of chronic exposure are based on projections from acute exposure at Hiroshima and Nagasaki. There may be a threshold for harmless exposure because of the fact that some radiation occurs naturally in the environment, and we have evolved to process it. However, health problems, such as cancer, are clustered around areas with radioactive contamination.

Radioactive Elements

There are sixty radioactive elements—both man-made and naturally occurring. Looking at the periodic table, elements with an atomic number greater than eighty have radioactive isotopes. Elements with an atomic number greater than eighty-three *only* have radioactive isotopes. Elements with fewer than eighty protons, such as iodine (atomic number 53), can also have radioactive isotopes. Below are some radioactive elements, their sources, and their health effects.

In its stable form, iodine is an important part of our diet. It's found in seafood, seaweed, dairy products, and grain. Absorbed in the thyroid, it prevents enlargement of the gland (known as a goiter). Iodine is especially important during pregnancy and infancy, when deficiency can lead to mental handicaps. Like stable iodine, radioactive iodine is also absorbed in the thyroid, only instead of nourishing the body, it causes thyroid cancer. This problem was exacerbated around Chernobyl because children had an iodine deficiency in their diets. The risk of thyroid cancer can be lowered by avoiding contaminated food and milk after a nuclear event and by

Stable iodine, found in seaweed, seafood, and many other foods, is absorbed by the thyroid as a nutritious part of the diet. Radioactive iodine is also absorbed by the thyroid but causes thyroid cancer.

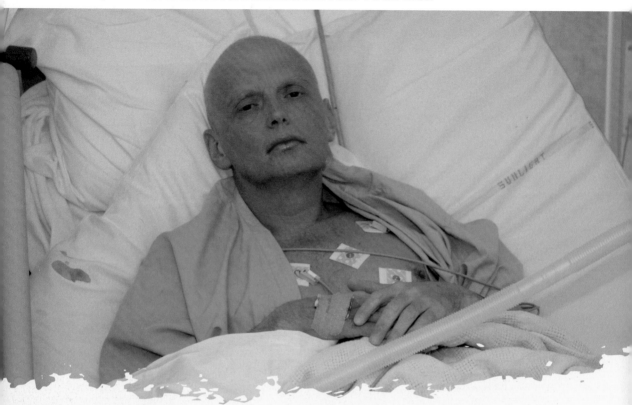

taking potassium iodide. This blocks the thyroid from absorbing the radioactive iodide. In Poland, potassium iodide was given to children following the Chernobyl incident and helped prevent thyroid cancer. In Belarus and Ukraine, children were not given the treatment, and the incidence of thyroid cancer increased. (Children and teens need more calcium than grownups because their bones are still developing, which means they must drink more milk. The Chernobyl disaster, sadly, poisoned the very nourishment children needed).

Alexander Litvinenko, a former Soviet KGB agent who had moved to London, was treated on November 20, 2006, for poisoning by polonium 210, an extremely deadly radioactive isotope. He died a few days later and is believed to have been murdered.

WHAT TO DO IN CASE OF NUCLEAR POWER PLANT DISASTER

Power plant meltdowns are extremely rare. If you are in the vicinity of one, you'll be directed to stay inside and take precautions such as closing all ventilation until it is safe to evacuate. Personal decontamination after a power plant disaster is surprisingly simple. You remove clothing and wash radioactive dust off your body with soap. If the area is contaminated, evacuation will occur. Then it's important not to eat or drink contaminated products, including milk from cows grazing in contaminated fields. Cleanup of the area, however, is expensive and time consuming. Residents may not be able to return home for many years, if at all.

Plutonium is heavier than iodine, so after a nuclear power plant disaster, the wind doesn't carry it as far. Also, unlike iodine, the danger of plutonium isn't ingestion. It actually passes through the digestive system, and only .04 percent is absorbed. But when inhaled, plutonium is absorbed by the lungs and moved into the bloodstream. It then affects all organs, particularly the liver and lungs, causing cancer. Plutonium exists in the soil primarily because of nuclear weapons testing. It can also contaminate the soil after a nuclear meltdown. Cesium 137 contaminated the regions surrounding Chernobyl and Fukushima. Cesium 137 is water soluble and is absorbed strongly by muscle tissues, leading to cancer.

Radioactive strontium and radium share chemical properties with calcium (radium is over a million times more radioactive than uranium). For this reason, they are absorbed in the bones and can cause tooth decay and bone cancer. A sad example of this occurred in the 1920s. The Radium Dial Company in Illinois and the United States Radium Company in New Jersey employed young women to paint glow-in-the dark watches. The paint contained radium. Though the companies knew of the health risks, they didn't alert the employees and even encouraged them to shape the paintbrushes with their mouths. The women suffered from toothaches and bone fractures. They developed tumors and blood diseases. The sufferers all died at a very young age. Radioactive strontium exists in the soil because of fallout from nuclear weapons testing and the Chernobyl meltdown.

Radon is the gas radium produces during radioactive decay. It occurs naturally in the soil because uranium, also in soil, breaks down into radium. Radon seeps through homes' foundations and mingles with the air. People then breathe in radon. At a certain level, radon can cause cancer. In fact it's the second-leading cause of lung cancer in America (after smoking), killing approximately fifteen thousand to twenty thousand people per year.

Tritium is a radioactive form of hydrogen. It occurs naturally in the air and water at levels that presumably don't cause health risks. It's also a byproduct of nuclear power plants. High doses could increase your risk of cancer. There have been tritium leaks at power plants,

but according to the Nuclear Regulatory Commission, they haven't posed a risk to the public.

Americium exists in the environment because of nuclear weapons testing and nuclear incidents such as Chernobyl. It's also used in industry. In fact, you probably have americium in your home; small amounts of americium 241, which results when plutonium 241 decays, are in smoke detectors. For that reason, you shouldn't dismantle your smoke detector, or, obviously, throw it in the fireplace. If ingested or inhaled, Americium is mainly absorbed in bone, muscle, and the liver.

A worker at a clock factory paints numbers with glow-in-the-dark paint, made so by radioactive radium 226. Many workers died from exposure to this paint.

4 RADIOACTIVITY: WHAT CAN YOU DO?

Understanding radiation and nuclear energy is the first step to addressing its positive and negative qualities. You now know that some radiation occurs in nature and household products. Most, such as radiation from a television or smoke alarm, is harmless. But your home should be tested for radon, which can reach dangerous levels. You've also learned that radiation is sometimes necessary, as in the case of a cancer treatment. Finally, you learned the difference between nuclear power and nuclear weapons, and positive and negative qualities of nuclear power. You also learned that radioactive contamination can have serious health consequences, including many types of cancer.

It's important to stay informed. Take enough science courses to understand safety issues, clean energy, and basic medicine. Or better yet, pursue a career in science. Read the newspaper for current instances of contamination. Check the EPA Web site to learn the history of your area and to find out

whether there is radioactive contamination. Go to community meetings in which discussions pertaining to nuclear power plants, radioactive waste, and cleanup efforts take place. Ask questions and advocate for your community's environmental health.

Form opinions based on fact. Decide how you feel about nuclear power, based on the pros and cons. Learn more about current sources of power such as coal and alternative, but less widespread, sources such as wind and solar energy. Think about what

Radon occurs naturally in homes, but high levels can be dangerous. You can test the levels in your home with a radon detection kit, shown here.

should be done with nuclear waste stored on-site at power plants. Also decide how you feel about nuclear weapons and testing.

When you're old enough, register to vote. Our elected officials make decisions on weapons, weapons testing, foreign policy, and energy policy. Often, during campaigns, we hear about "hot button" issues, such as taxes, jobs, and health care. Those are important issues. But if you're interested in energy and foreign policy, dig d eeper to learn the candidates' stand on these topics. You can also learn more about government policies and current events by participating in debate and mock United Nations events.

Finally, when you enter the workplace, handle any radioactive materials with care. Manufacturing, medical, scientific research,

Students at Boston University register to vote. Voting is one way to make your voice heard on issues such as defense, environmental protection, and nuclear power.

energy, and military jobs may require you to handle radioactive elements. It's important to know the proper safety procedures, what to do in case of an emergency, and how to report unsafe practices. Never assume that because someone outranks you, he or she has all the answers. We each must voice concerns about unsafe practices and faulty decision making.

Others like you have voiced their opinions for or against nuclear power. Environmentalists are split on the issue. Greenpeace, an environmental action group, has protested the existence of nuclear power plants, stating that another Fukushima-type disaster could occur any time. The Environmental Defense Fund, however, says that to combat global warming, nuclear power must be considered. President Barack Obama supports nuclear power, even after the Fukushima meltdown. However, he said that we must learn from that disaster.

People also disagree on the topic of nuclear disarmament. Some believe nuclear bombs are a necessary evil to prevent a third world war (because of the threat of devastation) and to maintain defense against countries that have bombs. On the flip side, people argue that, should a third world war occur, so will the very destruction that nuclear bombs were supposed to prevent.

Cleaning Up the Past

Others are tackling radioactive contamination that has already occurred. The EPA has a number of radioactive sites on the national priorities list, including seven each in California and Pennsylvania, six in New Jersey, five in Colorado, and four each in Florida, Missouri, Washington, and New York. Many are the results of government testing; others are related to industry.

Both government and industry now follow stricter regulations for handling radioactive materials and storing waste. In industry, the orphan sources program is one example. Through the years, power plants have also become safer. After the partial meltdown at Three Mile Island, safety measures and oversight were increased in the United States. No deaths occurred, and the average residents' exposure to leaked radiation appears to be less than a single chest X-ray. But at the time, authorities feared the worst: the release of large amounts of radiation. Afterward, workers received better training. Emergency procedures were revamped. Inspectors and regulators became more hands-on. The design of power plants also improved.

After the September 11, 2011, terrorist attacks, safety at power plants was increased to prevent terrorists from entering plants. The Fukushima disaster raised concerns among the U.S. public about power plants and natural disasters. The NRC stated that power plants in the United States were built to withstand natural disasters. However, it formed a task force, which suggested improvements such as off-site backup energy.

Last, but definitely not least, the United Nations is working to stop the proliferation of nuclear weapons, which have caused the most deaths from radioactive contamination by far after the two attacks in Hiroshima and Nagasaki.

MYTHS & FACTS

Myth: All radiation is caused by human technology.

Fact: Some radiation is caused by radioactive elements in the soil and in cosmic rays.

Myth: All radiation should be avoided.

Fact: Radiation can be helpful in curing cancer. Some radiation, such as that encountered in X-rays, is minimal and worth the risk since diagnosing broken bones is important to treating them.

Myth: A nuclear bomb has never been set off.

Fact: The atom bomb dropped on Hiroshima and the hydrogen bomb dropped on Nagasaki, killing hundreds of thousands, were nuclear bombs. Today's nuclear bombs are even more powerful.

10 GREAT QUESTIONS TO ASK SOMEONE WHO WORKS FOR THE EPA

1. How do I calculate my yearly radiation exposure?

2. How can radiation both cause cancer and treat cancer?

3. How can I check the radon level in my home?

4. How should workers protect themselves when working with radioactive material?

5. What are the risks and benefits of nuclear power?

6. How does nuclear power compare to other power sources currently in use?

7. How safe are nuclear power plants in case of an attack or natural disaster?

8. What can be learned from the Fukushima Daiichi Nuclear Power Plant meltdown?

9. What is the environmental impact of a nuclear meltdown?

10. What other green energy sources could be used for power?

GLOSSARY

atom The smallest unit of an element. It contains a nucleus, with electrons revolving around it.

contamination The state of being polluted or radioactive because of exposure to a pollutant or radioactive material.

dirty bomb A traditional (nonnuclear) explosive containing radioactive materials.

fallout Nuclear particles that fall to the earth after a nuclear explosion.

meltdown The melting of fuel rods at a nuclear power plant, resulting in the release of radioactive materials.

nuclear bomb An explosive device that gives off a great burst of energy caused by nuclear fission or fusion and releasing radioactive particles into the air.

nuclear power Power generated by controlled nuclear fission.

nuclear reaction The process by which a nucleus and a subatomic particle collide to produce products different from the initial particles.

radiation Energy emitted in particles or waves (such as alpha and beta particles and gamma and X-rays) when an atom disintegrates or is split.

radioactive Emitting radiation.

FOR MORE INFORMATION

Canadian Nuclear Safety Commission

280 Slater Street

P.O. Box 1046, Station B

Ottawa, ON KIP 5S9

Canada

(800) 668-5284

Web site: http://www.nuclearsafety.gc.ca

The Canadian Nuclear Safety Commission regulates nuclear energy and materials.

Environmental Protection Agency (EPA)

Ariel Rios Building

1200 Pennsylvania Avenue NW

Washington, DC 20460

(202) 272-0167

Web site: http://www.epa.gov

The EPA monitors, investigates, and regulates environmental matters.

Greenpeace

702 H Street NW

Suite 300

Washington, DC 20001

(202) 462-1177

Web site: http://www.greenpeace.org

Greenpeace takes action for a healthy environment and opposes nuclear energy
.

U.S. Department of Energy

1000 Independence Avenue SW

Washington, DC 20585

(202) 586-5000

Web site: http://www.energy.gov

The U.S. Department of Energy works to achieve scientific and technological leadership for America in terms of energy. It also works toward nuclear security and to clean up the environmental aftermath of the Cold War.

U.S. Nuclear Regulatory Commission (NRC)

Washington, DC 20555-0001

(800) 368-5642

Web site: http://www.nrc.gov

The NRC regulates commercial uses of radioactive materials, such as nuclear power plants.

Web Sites

Due to the changing nature of Internet links, Rosen Publishing has developed an online list of Web sites related to the subject of this book. This site is updated regularly. Please use this link to access the list:

http://www.rosenlinks.com/IDE/Radio

FOR FURTHER READING

Bortz, Fred. *Meltdown! The Nuclear Disaster in Japan and Our Nuclear Future*. Minneapolis, MN: Lerner, 2012.

Cappacio, George. *Cancer Treatments* (Advances in Medicine). Pelham, NY: Benchmark, 2012.

Freese, Susan. *Nuclear Weapons* (Essential Issues). Minneapolis, MN: ABDO, 2011.

Friedman, Lauri. *Nuclear Power* (Introducing Issues with Opposing Viewpoints). Farmington Hills, MI: Greenhaven, 2009.

Hersey, John. *Hiroshima*. New York, NY: Knopf, 1985.

Karam, Andrew. *Radioactivity* (Science Foundations). New York, NY: Chelsea House, 2009.

Laughlin, Robert. *Powering the Future: How We Will (Eventually) Solve the Energy Crisis and Fuel the Civilization of Tomorrow*. New York, NY: Basic Books, 2011.

Mahaffey, James. *Radiation* (Nuclear Power). New York, NY: Facts On File, 2011.

Mara, Will. *The Chernobyl Disaster: Legacy and Impact on the Future of Nuclear Energy*. Pelham, NY: Benchmark, 2010.

Miller, Connie Colwell. *Marie Curie and Radioactivity* (Graphic Library: Inventions and Discovery). Minneapolis, MN: Capstone, 2007.

Sullivan, Edward. *The Ultimate Weapon: The Race to Develop the Atomic Bomb*. New York, NY: Holiday House, 2007.

BIBLIOGRAPHY

Hargreaves, Steve. "Nuclear Waste: Back to Yucca Mountain?"
CNN, July 11, 2011. Retrieved Feb. 17, 2012 (http://money.cnn.
com/2011/07/06/news/economy/nuclear_waste/index.htm).

Hersey, John. *Hiroshima*. New York, NY: Knopf, 1985.

Inskeep, Steve. "Fear Dominates Discussions on Nuclear Power."
NPR, March 22, 2011. Retrieved Feb. 18, 2012 (http://www
.npr.org/2011/03/22/134755650/Fear-Stokes-Discussions-On
-Nuclear-Power).

Nobel Prize. "The Development and Proliferation of Nuclear
Weapons." Retrieved Feb. 5, 2012 (http://www.nobelprize.org/
educational/peace/nuclear_weapons/readmore.html).

PBS.org. "The Karen Silkwood Story." Retrieved Feb. 20, 2012
(http://www.pbs.org/wgbh/pages/frontline/shows/reaction/
interact/silkwood.html).

Read, Piers Paul. *Ablaze: the Story of the Heroes and Victims of
Chernobyl*. New York, NY: Random House, 1993.

Roan, Shari. "Possible Health Effects of Nuclear Crisis." *Los Angeles
Times*, March 16, 2011. Retrieved March 5, 2012 (http://articles.
latimes.com/2011/mar/16/health/la-he-japan-quake-radiation
-20110316).

Windridge, Melanie. "Fear of Nuclear Power Is Out of All
Proportion to the Actual Risks." *Guardian*, April 4, 2011.
Retrieved Feb. 18, 2012 (http://www.guardian.co.uk/science/
blog/2011/apr/04/fear-nuclear-power-fukushima-risks).